Santo Armenia

Historical reflections of Physics:

From Archimedes, …, Einstein till present.

Preface by
Carmelo Vindigni

Translated by
andreacrok

1

To my family: to my daughter Gabriella, to my son Pietro, to my daughter Marta, who is not a child anymore, and to my wife Marinella.

To Carmelo Vindigni.

1ˢᵗ edition: September 2019

Index

By CARMELO VINDIGNI

For this forth book, which you as well present as "This is the fourth work of a tetralogy, right now demanded, which gives birth to the pursuit of the Truth", you sked me to be the author of the Preface once again.

It is with renewed strength and pride that I pursue the task you gave me.

Ausculting your shell, Odysseus, you embarked on your historical adventure to Knowledge. You observed and saw that despite the progress, as a matter of fact, the mistake made by Archimedes, by Stevin and by Galileo due to the lack of instinctive understanding, has been afterwards kept and exacerbated by others: Newton, Lagrange, Einstein and The Scientific Community.

From your ceaseless rowing and digging, culminating in your "*new salt march*", may a new scientific confrontation begin and lead to the path of Knowledge, ultimately transcending the millennial mistake of *considering the weight of the bodies unalterable.*

You have already announced your fifth work, focused on new fronts, and not satisfied with sailing and digging, now you desire to fly, too?

May Dedalus and Prometheus be your companions.

This is the fourth work of a tetralogy, right now demanded,
which gives birth to the pursuit of the Truth.
The first work is:

GALILEO AND EINSTEIN

Reflections on the general relativity theory

The free fall of bodies

The shape of solid bodies

The second work is:

Archimedes

Reflections on buoyant bodies

The shape of solid bodies

The third work is:

ARCHIMEDES – GALILEO – NEWTON – EINSTEIN

The shape of solid bodies

The shape of fluid-filled containers

Unveiling – Epiphany

The experiments thus present are to be watched on YouTube: Armenia Santo or through the web site www.armeniasanto.it

Since the down of time philosophers considered the weight of bodies unalterable.

Euclid based his geometry on the fundamental entities (point, line and plane) and on postulates. One above all: between two points passes one and only one straight line. Geometry is a human construction, non-existing in the natural world.

Archimedes developed his studies in two fields:
1. geometry;
2. nature.

The geometry field includes the evaluation of the perimeter of flat shapes, the evaluation of the surface of flat shapes, the evaluation of the volume of solid shapes, the assessment of the barycentre of flat and solid shapes, etc.

The natural field includes the balance of bodies, the study of the buoyancy phenomenon etc.

Archimedes, confident of the sufficient reason principle or due to a spontaneous knowledge, developed his studies on nature basing it on postulates.

Modern and contemporary Physics (classical, relativistic and quantum), from Galileo till present, despite the definition of being experimental, are still built on postulates based on which theory after theory is constructed.

In this study the millennial mistake will be underline, an error derived from the postulate of considering the weight of bodies unalterable, even though it was outdated by Newton's universal law of gravity, based on which the weight of bodies changes with their position (variation of distance from the centre of the earth), as a matter of fact it persists and with the passing of time it exacerbates.

This millennial mistake includes at present three of them:
1 - Newton himself considered the weight of bodies unchangeable with their shape;
2 - about forty years ago the Scientific Community established that scales (with equal arms, analogues and digital) do not measure the weight but the mass of bodies;
3 - from the 20th May 2019 the Scientific Community, looking to redefine the fundamental physical quantities, hasn't established the standard sample nor the fabric or the sizeable shape for the kg mass.

Once the detailed knowledge of Kebble's Scale (used to redefine the kg per mass) will be spread publicly, it will be known if there is or there isn't a forth mistake, depending on the regular or the irregular shape of the system (the regular shape is achieved as a mere coincidence and not as a willing choice).

The historical era is divided into three periods:
1 - ancient, from Archimedes to Galileo;
2 - modern, from Newton to Lorànd Eötvös;
3 - contemporary, from Einstein till present.

From the historical research on the present study, news insights emerged:
a)   the study of buoyant bodies;
b)   the study of hanging/rolling bodies.

Chapter I. General Ideas

1.1 Body Weight, gravity weight, mass, gravity law

The concepts of mass and of gravitational pull didn't exist before Newton. Therefore, from Archimedes to Galileo the so-called bodies, also called weights, have a given quantity of physical matter and a peculiar body weight equal to the weight gravity (heaviness); from experience, this body weight is considered unaltered, unchanging, constant.

Newton introduced the idea of mass and formulated the universal law of gravitation. Moreover, he defined the weight of a body as the gravitational force that the Earth exerts on a body with a certain mass (there is no distinction between inertial mass and gravitational mass). The gravitational pull exerted between two bodies is inversely proportional to the square root of their distance. Thus, from Newton on a body has a certain mass and a correspondent weight. Each body consist of a certain mass which is unalterable, while its weight changes with its geographical coordinates: latitude, longitude and altitude. If a body was taken from the Earth to the Moon, or the other way around, its mass wouldn't change, it's its weight which changes, instead.

After comparing the recalled ideas, we have:
a) the term weight, within the meaning used before Archimedes till the present day, refers to two different concepts; the first one is the amount of substance (unchanging), the second one is the gravity pull (changing);

b) the ancient term weight gravity (once considered unalterable) is linked to the gravitational pull (afterwards considered variable);
c) the ancient term weight gravity, which was used by Galileo in the meaning of gravity but was considered unalterable; from now on in this book I will only use the term weight gravity referring to the scientists prior to Newton.

1.2 Barycentre, Centre of weight gravity, centre of gravity, centre of mass

The barycentre is one among the many centre points of a geometric figure; this is valid at any point of history.

The centre of a falling body, when we refer to ancient times (from Archimedes to Galileo), its definition is slightly different for each author (Archimedes, Pappus, Stevin), is the centre point such that if a body was hanging from it, the body wouldn't lose balance.

The centre of weight gravity, when we refer to modern and contemporary ages, is the point where the gravitational force (weight) is applied; it is also called barycentre.

The centre of mass, during modern and contemporary times, is the point in which the static moment of every axis passing through it is zero. In this peculiar case in which the distribution of the mass is uniform, the centre of mass corresponds to the geometrical barycentre of the figure considered; thus, the reference axis is called barycentric axis.

During the ancient time, since the weight of bodies was considered unalterable, the centre of weight gravity used to match the geometrical barycentre.

In the peculiar case of uniform mass distribution, it is underlined that the centre of mass matches with what was called centre of weight gravity during ancient times.

During modern and contemporary times, even if it was known that the gravity force (weight) of a body was alterable, how could they name it centre of gravity or "barycentre"? Moreover, how could they believe that the centre of mass corresponded to the centre of gravity?

The centre of mass corresponds to the geometrical barycentre exclusively when the mass distribution is uniform.

If we treat the simple case of uniform mass distribution, also when there is a symmetry, the centre of gravity never matches with the centre of mass. The classical example is the circle in which the centre of mass corresponds to the centre of the circle, but the centre of gravity lies beneath it and therefore cannot be the barycentre.

The rate of distance between the centre of gravity and the centre of mass peaks on the earth's surface, diminishing with altitude. Moreover, it increases with:
a) mass growth;
b) density reduction;
c) shape, growing from the sphere to the cylinder to the cube.

All of this is the consequence of my scientific discovery: "*The shape of solid bodies*".

"*The love for the Pursuit of the Truth and Knowledge*", *regardless of "The shape of solid bodies",* has led me to new insights about:
a) buoyant bodies;
b) hanging/rotating bodies.

2. Shape of solid bodies

Newton formulated the universal law of gravity referring to bodies as point masses: this model is the one adopted nowadays.

With this setting, when calculating the weight of a lot of objects (equal in mass and fabric but different in shapes; equal in fabric and shape but different in mass; different in mass, equal or different in fabric, equal or different in shape) all of them placed on the earth's surface or on a reference plane (for example one side of a scale), nowadays it is assessed that they have zero distance both from the earth's surface and from the reference plane. Knowing this, two bodies equal in mass (notwithstanding the shape nor the fabric) must have the same weight. Moreover, two bodies with different but proportional masses (notwithstanding the fabric not the shape) must have proportional weights.

Every author from Archimedes to Galileo, knowing the postulate of the inalterability of the weight gravity, claimed this.

This is odd since it has been claimed by every author, from Newton till present, despite knowing that the gravitational pull changes with altitude.

Moved by The Love for the Pursuit of the Truth and Knowledge, knowing the Socrates' Art, I revealed that not each body could be seen as a point mass. Instead, each body must have its distance from the centre of mass of the earth's surface or from the reference plane: *nature itself takes this into account.*

This is my scientific discovery:
The shape of solid bodies

And its physical law is:

In every natural phenomenon observed, in every experiment carried out, when there is a gravitational pull, the shape of solid bodies needs to be considered.

This law is valid for my scientific discovery due to two reasons:
1) it is the consequence of a correct application of Newton's universal law of gravity; on one side, it continues to refer to the point mass, on the other side, it takes into account the distance between a body and the centre of mass of the earth's surface or a reference plane, so as to characterize for each body its reality, "*matter and shape*", and to obtain the closest behaviour to the natural one; This aspect had not yet been considered in Physics;
2) it is confirmed by the experiments on:
   a) the weight of bodies in different positions of a scale pan;
   b) the immersion of bodies in water with differ position.

    The relevance of this scientific discovery is double:
1) analytic, because it should be integrated and adjusted with Newton's universal law of gravitation;
2) experimental, because its effect (variation of weight with the different position of a body on a scale pan) is visible also with ordinary scales.

If Archimedes, with his scales, had searched for this effect, would he have found it?

It is with renewed love for the Pursuit of the Truth and Knowledge that I found out this effect in water, a "*natural scale*", with experiments about buoyancy; I observed that each body with a density similar to the density of water, changes their state (floatation, suspension, sinking) when the shape of the body changes (the gravitational shape changes notwithstanding the geometrical shape) when it is placed in water.

If Archimedes, with his theorical model, had searched for this effect, would he have found it? No!
With his theorical model is impossible without experiments and assuming the postulate of the weight gravity.
Archimedes should have made experiments to see it.
Did Archimedes experiment?
Did Archimedes experiment but didn't see it?

It is impossible to know.

Why has this effect never been seen even afterwards by Stevin, Galileo and modern scientists? … It is odd, because most part of them, Stevin and Galileo, made experiments and scientists are currently experimenting, too.

Is it possible that somebody experimented but couldn't see it in the past?
We will never know.

Nonetheless, the story of the fresh egg floating, and the old egg sinking was known to all of them.

I hadn't known about it until last year, when I saw a video on YouTube and I was astonished. How is it possible that nobody has ever considered this experimental evidence?

With the passing of time or adding salt into the water, the fresh egg goes from floating to sinking, but in the middle, it passes through every position of balance *"in reality"* at every depth and not through the only position allowed by *"the theorical model"*.

The effect of *"The shape of solid bodies"* must be acknowledged by everyone: from children to adults. Through this new proper vision, the common sense must change to pursue the Pursuit of the Truth ad Knowledge.

Chapter III Ancient: from Archimedes to Galileo

3.1 Archimedes

Archimedes led his studies on nature in two fields, the balance of bodies and the buoyancy of bodies, always separating them.

3.1.1 *Body balance*

The studies of the lever, of the balance of resting bodies and of the balance of hanging bodies, followed the two postulates, from Archimedes on:
a) The weight gravity of bodies is unalterable;
b) The directions of weight bodies are parallel lines.

3.1.2 *On buoyant bodies*

From Archimedes on, the study of buoyant bodies followed the postulate:
"Given a liquid with such properties that in its consecutive and equally placed portions, the less compressed one is pushed by the most compressed one and that each one of its parts is compressed by the perpendicular of the liquid pushing from above, provided that the liquid itself is included in something and compressed by something else".

Therefore, he demonstrated the first two propositions:

-    First proposition: *"If a surface is cut by a plane through a point which remains the same, it generates a circumference and since it always has the centre in the point where the plane is cut, the surface obtained is a sphere".*

- Second proposition: *"Each liquid assumed motionless and quiet, takes the form of a sphere centred on the centre of the Earth".*

The consequence is that the direction of the weight gravity of bodies is radial.

3.1.3. *Comparison between two studies*

The consequences of the comparison between two very different studies are:
a) the weight gravity of bodies is unalterable;
b) the weight gravity of bodies has the direction of a straight parallel line (for the *stability of bodies*) and of a radial line (for *buoyant bodies*).

3.1.3.1 *Lever*

The postulate for the lever is: *two bodies in bilateral symmetry are balanced.*

This postulate, which assumes the previous two postulates (the weight gravity is unalterable; the directions of the weight gravity are straight parallel lines), had been shared by many author's previously but it is no longer considered valid because:
a) it had already been contradictory because of the weight gravity directions; one thing is using it for practical reasons and one thing is using it with didactic purposes and theorical-scientific purposes;
b) the postulate "the weight gravity of bodies is unalterable" is no longer valid.

A detailed study has been dealt with in the 3rd Work.

3.1.3.2 *Resting bodies*

The study of the stability of resting bodies in equilibrium (for example, a body placed on the vertex of a concave upward parabola: when the body is moved from the vertex, it goes back to it), instable in equilibrium (for example, a body placed in the vertex of a concave downward parabola: when the body is moved from the vertex it departs from it) and indifferent in equilibrium (for example, a body lying on an horizontal plane: when the body is moved it remains stable in every position), had been shared by many authors but it is only valid for bodies stable and instable in equilibrium.
The indifference in equilibrium, instead, doesn't exist in the natural world. In fact, due to the radial direction of the weight gravity, the weight force is divided into two components:
a) one is perpendicular to the resting surface, which has no effect as it is balanced by the supporting force;
b) the other is parallel to the horizontal resting surface, whose direction is towards the initial position, and it affects the stability.

Thus, a new concept is introduced, "*rate of stability of resting bodies*", for bodies with balanced equilibrium the result is:
a) maximum rate of displacement on the vertical plane, going upwards;
b) medium rate of displacement on the inclined plane, going upwards;
c) minimum rate of displacement on the horizontal plane.

The shape of solid bodies is influential.
After fixing the body mass and the vertical displacement from the centre of the mass, the shape determining the maximum value of the "*rate of stability of resting bodies*"

is, in decreasing order, the cube, the equilateral cylinder and the sphere. This is because their weights have:

$$P_{cube} > P_{equilateral\ cylinder} > P_{sphere}.$$

Likewise, introducing a new concept, "*rate of stability of resting bodies*", for the instable equilibrium the result is:
a) Maximum rate of displacement causing the vertical free fall of bodies;
b) Medium rate of displacement on the inclined plane, going downwards, with a decreasing rate tending toward zero when the inclined plane tends toward the horizontal plane.

The shape of solid bodies is influential.
After fixing the body mass and the vertical displacement "h" from the centre of the mass, the shape determining the maximum value of "*the rate of stability of resting bodies*" is, in decreasing order, the cube, the equilateral cylinder and the sphere. This is because their weights have:

$$P_{cube} > P_{equilateral\ cylinder} > P_{sphere}.$$

3.1.3.3.   *Hanging bodies*

The study of the stability of hanging bodies in equilibrium (for example, with the suspension point placed above the centre of weight gravity, which corresponds to the centre of mass: the rotating body goes back to its initial position), instable in equilibrium (for example, with the suspension point placed below the centre of weight gravity, which corresponds to the centre of mass: the rotating body departs from it) and indifferent in equilibrium (for example, with the suspension point placed below the centre of weight gravity, which corresponds to the centre of mass: the rotating body is stable in every position),had been shared by many authors but it is only valid for bodies stable and instable in equilibrium.

The indifference in equilibrium, instead, doesn't exist in the natural world, due to the radial direction of the weight gravity and most of all due to the variation of the gravitational pull. Given that the centre of gravity doesn't correspond to the centre of mass, the weight force gives rise to a moment regarding the centre mass, thus the body is not stable in every position.

The detailed study of the displacement of a free body rotating around its centre mass will be discussed in another work.

I will only anticipate that I saw intuitively that the study is built on *"the geometry of masses"*, concerning the two most important moments of inertia. A new concept is also introduced for this new study, the *"rate of stability of hanging bodies"* for balanced stability and the *"rate of instability of hanging bodies"* for unbalanced stability.

3.1.3.4 *Buoyant bodies*

With the passing of time, a lot of authors have studied buoyant bodies, such as Stevin and Galileo whose results are valid nowadays too; on the other hand, there is the conviction that Archimedes' results remained unaltered and so they match with Stevin's and Galileo's.

Archimedes' propositions are indicated below:

- Third proposition: *"Solid bodies with the same weight of the liquid, if submerged in it, dive in without descending below the surface of the liquid nor plunging to the bottom"*;

- Forth proposition: *"Of two solid bodies, the one lighter than the liquid and submerged into it, doesn't immerge entirely but a part of it comes to the surface of the liquid"*;

- Fifth proposition: *"Of two solid bodies, the one lighter than the liquid dives into it to have a correspondence between the volume of the liquid, for the solid part submerged, and the weight of the entire solid body"*;

- Sixth proposition: *"Solid bodies, lighter than the liquid, when pushed inside of it experiences an upward force with an intensity proportional to their volume"*;

- Seventh proposition: *"Solid bodies heavier than the liquid they are submerged in, plunge to the bottom and their weight inside the liquid decreases of a quantity corresponding to the liquid displaced, equal in volume to the solid body"*.

This theorical model and those of Stevin and Galileo, are revealed after my experiments.

The postulate *"two bodies in bilateral symmetry are balanced"* concerning the lever is erroneously derived from experience, *"intuitive knowledge"*; From which experience is the theorical model of buoyant bodies derived?

Moreover, it is also incomplete and contradictory, in particular:

1st QUESTION-OBSERVATION

What are the conclusions of the propositions for the theorical model (liquid sphere in a state of rest) worth when the Earth has completely different properties compared to the theorical model?

2nd QUESTION-OBSERVATION

If the theorical model is a "liquid sphere in a state of rest", what is the bottom of this "liquid sphere in a state of rest"?

3rd QUESTION-OBSERVATION

When we refer to the seventh proposition, solid bodies heavier than the liquid they are submerged in, but different from each other, to what bottom are they headed? How is the weight of those solid bodies decreasing in a differently compressed liquid?

4th QUESTION-OBSERVATION

When we refer to the sixth proposition, the difference between Archimedes' push and the body weight is zero; Is the obvious conclusion for one of the many conditions of stability below the surface of the liquid, "*body submerged in water*", missing? Could the theorical model predict this effect?

5th QUESTION-OBSERVATION

The statements of the III, IV, V, VI and VII propositions were made with respect to "every solid body", hence without considering their shape; their demonstration was nevertheless made considering the "pyramid" which is a shape with a vertical symmetry.

Thus, with the limitations of a theorical model, is their demonstration valid?

6th QUESTION-OBSERVATION

Why should the proposition of the theorical model be valid for a liquid immersed in a container, when the postulate states "*...provided that the liquid itself is included in something and compressed by something else*"?

3.1.4 *Conclusions*

Experiments are crucial to understand nature better. Archimedes developed his studies with the help of theorical models derived from a postulate without even experimenting, thus giving birth to severe mistakes. For his time the entity of his mistakes was approximated with a good accuracy, despite the uncertainty in the behaviour of bodies below the surface of the liquid that do not touch the bottom.

We will see in the following pages that, during modern times, scientists have built other principles based on these mistakes (Galileo's principle of inertia) and they also have built new theories (Einstein's relativity theory), adding more and more mistakes.

Despite the technological and scientific development, it almost looks as if two millennia have passed in vain: at present, the weight of a body is still addled for its mass, thus adding more severe mistakes.

3.2 Stevin

Stevin kept his two natural studies on the stability of bodies and on buoyant bodies separated.

3.2.1 *Body balance*

Stevin's studies on the lever, on the balance of resting bodies and on hanging bodies were based on the postulates:
a) the weight gravity is unalterable;
b) the directions of weight bodies are parallel lines.

3.2.2 *On buoyant bodies*

Stevin's study of buoyant bodies was carried out holding to be true that liquids are incompressible (they have a homogeneous weight gravity).
The directions of the weight gravity are parallel lines.

3.2.3 *Comparison between two studies*

The consequences of the comparison between two very different studies are:
a) the weight gravity of bodies is unalterable;
b) the directions of the weight gravity of bodies are always parallel lines.

3.2.3.1 *Lever*

Stevin preserves Archimedes' first postulate on the lever, *"two bodies with a bilateral symmetry are balanced"*. On the other hand, he doesn't accept the second part of the postulate which says: *"there is a balance even when bodies are inversely proportional to their pans"*, and he demonstrates it.

This choice (refusing to accept the second part of the postulate on the lever and demonstrating it) is an open critic to Archimedes, who based his studies on postulates.

This behaviour, however, show a great maieutic contradiction; he demonstrates the second part of the postulate about the lever basing his study on the first postulate.

For Stevin as well as Archimedes, the first postulate and the demonstration of the second part, which are based on

the previous postulates (the weight gravity is unalterable; the directions of the weight gravity of bodies are always parallel lines), is no longer considered valid for two reasons:
a) It had already been contradictory because of the weight gravity directions; one thing is using it for practical reasons and one thing is using it with didactic purposes and theorical-scientific purposes; it is also valid for Archimedes;
b) The postulate "the weight gravity of bodies is unalterable" is no longer valid.

A detailed study has been dealt with in the 3rd Work.

3.2.3.2 *Resting bodies*

Stevin's results on the study of resting bodies are equal to Archimedes'.

3.2.3.3 *Hanging bodies*

Stevin's results on the study of resting bodies are equal to Archimedes'.

3.2.3.4 *Buoyant bodies*

Stevin's study on buoyant bodies was carried out theoretically supposing an incompressible liquid (with a homogeneous weight gravity), placed in a container where he hypothesizes a "superficial vase" to replace the liquid inside the vase with any other body.

From the comparison between the two specific weight gravities, the one of the liquid (gr_l) and the one of the body (gr_c), it is demonstrated that three possible states exist:
a) **gr_l > gr_c** buoyant body (part of the body is immersed in the liquid and part of it is in the air);

b) **gr$_l$ < gr$_c$** sunk body (body entirely immersed in the liquid and touching the bottom of the container with its base);

c) **gr$_l$ = gr$_c$** body in indifferent equilibrium (body placed below the free surface of the liquid, no matter the depth, without touching the bottom).

The buoyancy state (**gr$_l$ > gr$_c$**) and the state with the sunk body (**gr$_l$ < gr$_c$**) are experienced in the everyday life: those two principles are in line with the natural ones.

The state of indifferent equilibrium (**gr$_l$ = gr$_c$**), "body placed below the free surface of the liquid, *no matter the depth*, without touching the bottom", conflicts with Archimedes' theorical principle which did not include the possibility for the body of being in every position "*no matter the depth*".

To be clearer, I rewrite Archimedes' third proposition:
- Third proposition: "*Solid bodies with the same weight of the liquid, if submerged in it, dive in without descending below the surface of the liquid nor plunging to the bottom*".

This conflict (repeated also with Galileo) is of vital historical importance because up until now, as written in academic textbooks (some of which were written by Stevin and some others by Galileo), nobody had seen it before.

Moreover, the theorical forecast of indifferent equilibrium formulated by Stevin and Galileo has never been confirmed by any experiment.

In my 2nd and 3rd works I demonstrated that Archimedes' third preposition is correct but incomplete, because it doesn't consider every other infinite position of stability at every depth; instead, Stevin's and Galileo's statements are incorrect.

Stevin's mistake is the consequence of an incorrect postulate, for which liquids have a *"uniform weight gravity"*.

My experiments in water demonstrate that the density of liquids is not constant but increases with depth. Moreover, in my 3rd work I underlined that this variation in density relies on the shape of the container; the associated experimental confirmation needs to be done with uncommon containers.

A detailed study of buoyant bodies, aiming to study the buoyant state and bodies hanging in water, will be discussed in another work hereinafter.

3.3 Galileo

Galileo kept his two natural studies (on the stability of bodies and on buoyant bodies) separate.

3.3.1 Body balance

Galileo's studies on the lever, on the balance of resting bodies and on hanging bodies were based on the postulates:
a) the weight gravity is unalterable;
b) the directions of weight bodies are parallel lines.

3.3.2 *On buoyant bodies*

Galileo's study of buoyant bodies was carried out holding to be true that liquids are incompressible (they have a homogeneous weight gravity).

The directions of the weight gravity are parallel lines.

3.3.3 *Comparison between two studies*

The consequences of the comparison between two very different studies are:
a) the weight gravity of bodies is unalterable;
b) the directions of the weight gravity of bodies are always parallel lines.

3.3.3.1 *Lever*

Galileo, like Stevin, preserves Archimedes' first postulate on the lever, "*two bodies with a bilateral symmetry are balanced*". On the other hand, he doesn't accept the second part of the postulate which says: "*there is a balance even when bodies are inversely proportional to their pans*", and he demonstrates it.

This choice (refusing to accept the second part of the postulate on the lever and demonstrating it) is an open critic to Archimedes, who based his studies on postulates.

This behaviour, however, show a great maieutic contradiction; he demonstrates the second part of the postulate about the lever basing his study on the first postulate.

For Galileo as well as Stevin and Archimedes, the first postulate and the demonstration of the second part, which are based on the previous postulates (the weight gravity is unalterable; the directions of the weight gravity of bodies are always parallel lines), is no longer considered valid for two reasons:
a) it had already been contradictory because of the weight gravity directions; one thing is using it for practical reasons and one thing is using it with didactic purposes and theorical-scientific purposes; it is valid also for Archimedes;

b) the postulate "the weight gravity of bodies is unalterable" is no longer valid.

A detailed study has been dealt with in the 3rd Work.

3.3.3.2 *Resting bodies*

Galileo's results on the study of resting bodies are equal to Archimedes'.

3.3.3.3 *Hanging bodies*

Galileo's results on the study of resting bodies are equal to Archimedes'.

3.3.3.4 *Buoyant bodies*

Galileo's study on buoyant bodies was carried out theoretically supposing an incompressible liquid (with a homogeneous weight gravity), placed in a container with vertical walls, the shape of the body could be a "cylinder" or "prism".

In line with Stevin, from the comparison between the two specific weight gravities, the one of the liquid (gri) and the one of the body (grc), it is demonstrated that three possible states exist:
a) **$gr_l > gr_c$** buoyant body (part of the body is immersed in the liquid and part of it is in the air);
b) **$gr_l < gr_c$** sunk body (body entirely immersed in the liquid and touching the bottom of the container with its base);
c) **$gr_l = gr_c$** body in indifferent equilibrium (body placed below the free surface of the liquid, no matter the depth, without touching the bottom).

The buoyancy state ($gr_l > gr_c$) and the state with the sunk body ($gr_l < gr_c$) are experienced in the everyday life: those two principles are in line with the natural ones.

The state of indifferent equilibrium ($gr_l = gr_c$), "body placed below the free surface of the liquid, *no matter the depth*, without touching the bottom", conflicts with Archimedes' theorical principle which did not include the possibility for the body of being in every position "*no matter the depth*".

To be clearer, I rewrite Archimedes' third proposition once more:
-   Third proposition: *"Solid bodies with the same weight of the liquid, if submerged in it, dive in without descending below the surface of the liquid nor plunging to the bottom".*

This conflict (repeated also with Stevin) is of vital historical importance because up until now, as written in academic textbooks some of which were written by Stevin and some others by Galileo, nobody had seen it before.

Moreover, the theorical forecast of indifferent equilibrium formulated by Stevin and Galileo has never been confirmed by any experiment.

In my 2nd and 3rd works I demonstrated that Archimedes' third proposition is correct but incomplete, because it doesn't consider every other infinite position of stability at every depth; instead, Stevin's and Galileo's statements are incorrect.

Galileo's and Stevin's mistakes are the consequence of an incorrect postulate, for which liquids have a "uniform weight gravity".

My experiments in water demonstrate that the density of liquids is not constant but increases with depth. Moreover, in my 3rd work I underlined that this variation in density relies on the shape of the container; the associated experimental confirmation needs to be done with uncommon containers.

A detailed study of buoyant bodies, aiming to study the buoyant state and bodies hanging in water, will be discussed in another work hereinafter.

3.3.3.5 *Horizontal plane – Principle of inertia*

As we already know, see my 3rd work, knowing Archimedes' postulate "*the weight gravity directions (vertical) are parallel lines*", Galileo formulated the principle of inertia (1st principle of the dynamics):

"*If there is no net force on an object, then its velocity is constant. Either the object is at rest (if its velocity is equal to zero), or it moves with constant speed in a single direction.*"

This principle, shared by the scientific community till present just like the principle of indifferent equilibrium, doesn't exist in the natural world instead. In fact, due to the radial direction of the vertical, the weight gravity is broken up into two different parts:
a) the first, perpendicular to the horizontal plain of rest doesn't have an effect because it is balanced by the binding reaction;
b) the second, parallel to the horizontal plane of rest determines at first a delayed movement which will stop and it's due to its straight direction in the initial position, secondly it determines an accelerated motion of opposite direction.

3.3.3.6 *The shape of solid bodies*

Galileo repeatedly asked himself whether the shape of a body could change or not the weight gravity. He always reached to the conclusion that the shape of a body didn't change the weight gravity.

In particular, when talking about the study of buoyant bodies (DISCOURSE ON BODIES THAT STAY ATOP WATER, OR MOVE IN IT) the conclusion has always been: "*The difference in shape between a given solid and another is never, in anyway, the reason why the body does or doesn't go to the bottom or to the surface; so that a solid, for example, spherical in shape goes to the bottom or to the surface when immersed in water, I say that if it had every other shape, the same solid in the same liquid would go to the bottom or to the surface, nor could its movement be blocked or averted by the amplitude of its displacement or any other mutate.*".

4.1 Newton

Newton establishes the universal law of gravitation, based on which the gravity (weight) of bodies changes with their position (variation of the distance of the centre mass, for the body in question, from the centre of planet Earth). Moreover, Newton is the father of the infinitesimal calculus.

Nevertheless, Newton states erroneously that:
a) body weight and body mass are proportional; instead a body with a mass proportional "n times" to that of a given body, so as not to have the same distance from the centre mass of the given body, but at a further distance its weight isn't "n times" proportional to the weight of that of the given body but is slightly inferior;
b) the shape is not influential on the weight of a given body (except for the cube, for the equilateral cylinder and for the sphere); instead ,even if the geometrical shape doesn't change, when the position of the body resting on earth's surface changes (its gravitational shape is transformed), so does the distance from the centre mass of the body and subsequently the weight of the body itself;
c) the shape is not influential on the weight of the given body (except for the cube, for the equilateral cylinder and for the sphere); instead, even if the quantity of matter doesn't change with the different shape of bodies (thus changing the gravitational shape), so the distance from the centre mass of the body changes and subsequently the weight of the body itself.

These conclusions are collected in his work "*On the system of the world*" (third book):

Corollary 1.- *"So the weight of bodies does not rely on the shape or the aggregation. Because if they changed with the shape, they would be once bigger and once smaller depending on the variety of their shapes, with the same quantity of matter. This is, though against every experience";*

Corollary 2.- *"Every body placed around the Earth gravitate on it and its weight - equal in distance from the Earth's centre – directly proportional to its mass".*

All of this, without even saying a word about the effects of the shape on the rotational inertia, studied in another work hereinafter, thus Newton's law of gravitation cannot be considered universal:
1) the shape doesn't affect the inertial *"translation"* effects;
2) the shape doesn't affect the inertial *"rotational"* effects.

Nobody, until today, had ever seen this mistake which doesn't always have the same entity but, as seen in the 1st work, is maximum on Earth's surface.

Moreover, this mistake is bigger in bodies with:
a) equal mass and minor density;
b) equal mass and a shape which determines a further distance from the centre mass;
c) different mass, when the mass increases.

Newton, who did a lot of experiments, in particular to see the influence of the different nature of the matter on the inertial mass and the gravitational mass (so much wasted work), didn't study buoyant bodies. This study, instead, was the key to accepting the influence of the shape of solid bodies on the weight of the bodies themselves.

Newton, although he knew that the body weight needed to change with its position, in particular the altitude on the earth's surface which affects of the shape of solid bodies, a millennial mistake due to the intuitive knowledge of believing the weight gravity (or weight for Newton) of bodies unalterable, to which is added to the belief of considering the weight gravity unalterable and so confirming every mistaken conclusion made by Galileo.

Aware of the gravity law, and so conscious of the influence of the body weight on its shape, I examined nature, *"water a natural pan"*, and with the experiments made in the second work, I confirmed my scientific discovery *"the shape of solid bodies"* which I had already predicted with the help of maieutic in the first work.

4.2 Lagrange

4.2.1 *The lever*

Lagrange, as well as Galileo and Stevin, keeps Archimedes' first postulate *"two bodies in bilateral symmetry are stable"*. Instead, he refuses the second part of the postulate on the lever, *"there is a balance even when bodies are inversely proportional to their pans"*, and he demonstrates it.

This choice (refusing to accept the second part of the postulate on the lever and demonstrating it) is an open critic to Archimedes, who based his studies on postulates.

This behaviour, however, show a great maieutic contradiction; he demonstrates the second part of the postulate about the lever basing his study on the first postulate.

For Lagrange as well as Stevin, Galileo and Archimedes, the first postulate and the demonstration of the second part, which are based on the previous postulates (the weight gravity is unalterable; the directions of the weight gravity of bodies are always parallel lines), is no longer considered valid for two reasons:

a) it had already been contradictory because of the weight gravity directions; one thing is using it for practical reasons and one thing is using it with didactic purposes and theorical-scientific purposes; it is also valid for Archimedes;

b) the postulate "the weight gravity of bodies is unalterable" is no longer valid.

A detailed study has been dealt with in the 3rd Work.

4.2.2 *Resting bodies*

Lagrange's conclusions are equal to Archimedes'.

4.2.3 *Hanging bodies*

Lagrange's conclusions are equal to Archimedes'.

4.2.4 *Buoyant bodies*

Lagrange took Archimedes' and Stevin's studies on buoyant bodies, compared them and didn't see the contradiction expressed in this study, holding to be true that they reached the same conclusions.

So, Lagrange reached to the same conclusions as Stevin.

4.2.5 *Kilogram mass*

During the French Revolution, the French nominated a scientific commission in charge of establishing the unit of measurement of the fundamental physical quantities; many scientists like Lagrange, Laplace, Lavoisier etc. took part in it.

The commission declared the kg mass as the mass of a litre of water in standard conditions.

The shape of the litre wasn't given, so a variety of different samples could have been used (infinite in theory), which gave a different value for the given weight, except for the regular shapes (cube, equilateral cylinder and sphere).

So that a new experimental and practical mistake was added to Archimedes' millennial mistake and to Newton's analytic mistake. Moreover, each measure included the mistake of considering that a weighted body had only one value, instead, there were as many values as the positions on the scale pans.

Afterwards, scientists were convinced that every sample of 1 litre (1 dm^3) coincided with 1 kg mass. Thus, the definition in force until the 20th May 2019 was changed into the platinum-iridium mass of an equilateral cylinder of 3,9 cm in diameter, preserved in the International Office of Weights and Measures of Sevres, France: it was just a casualty that a regular shape was chose, "the equilateral cylinder".

Nonetheless, two mistakes were perpetuated:
a) samples used with all scales to be weighted aren't regular in shape;
b) all the measurements of the weighted body are thought to have only one value, when, instead, they have as many values as the possible positions on the pan.

4.3 Lorànd Eötvös

4.3.1 *Inertial mass, gravitational mass*

Loràrd Eötvös experimented using a *"torsion balance"* to assess a potential difference between the inertial mass and the gravitational mass. The experiment resulted in an equality with an initial accuracy of 5×10^{-9} and further improved to 3×10^{-14}. NASA, with a different experiment into the space, aims at reaching an accuracy of 1×10^{-18}.

Einstein undermined al those efforts, made from Newton on, with the help of his postulate about the equivalence principle (weak equivalence principle), which states that the inertial mass and the gravitational mass are equal. This concerns Einstein, but I discredited the general relativity theory under various aspects, in particular under my scientific discovery of *"the shape of bodies"*.

4.4 Mach

Mach, to whom I owe all my historical knowledge on mechanics and on the "psychology of research", didn't see anything despite his scrupulousness. By doing so, he paved the way to Einstein's axiomatic relativity theory, although he wanted a mechanic free of Newton's metaphysics. Afterwards, also due to his philosophic theory based on *"neutral monism"*, was criticized by many like Lenin and Einstein himself.

Chapter V Contemporary: from Einstein till present.

5.1 Einstein

5.1.1 *General relativity theory*

Einstein, implying his "special relativity", formulated his general relativity theory supposing that Galileo's principle on the free fall and that Archimedes' postulate (the directions of the weight gravity of bodies are straight parallel lines) were correct. Einstein used improperly Lorànd Eötvös' experience, aiming at evaluating the difference between the inertial mass and the gravitational mass instead of evaluating the truthfulness of Galileo's principle on the free fall of bodies (whose original version had been refused centuries earlier and stated that the gravitational acceleration is constant, equal for every body).

I refused the general relativity theory due to the following reasons:

a) it is inappropriate to build a contemporary theory, such as the relativity theory which aims at studying the entire universe, upon the assumption of Archimedes' millennial approximation (the directions of the weight gravity of bodies are straight parallel lines when, instead, they are radial); as a consequence, I demonstrated that the relative mistake doesn't go beyond the current approximation of $1x10^{-14}$; it is vital to take this into account;

b) I demonstrated that Galileo's principle on the free fall is incorrect, because every body undergoes its gravitational acceleration which is variable (see 1st work);

c) for the consequences of my scientific discovery *"the shape of solid bodies"*, so that there is no equivalence between the linear movement in a straight line and the gravitational field.

5.2 Scientific Community

5.2.1 *Weight and mass*

About 40 years ago the Scientific Community, (due to commercial reasons?) established that scales (with equal arms, analogues and digital) do not measure the weight but the mass of bodies.

By doing so, it was added to the pre-existing mistake on the measure of the weight, the erroneous belief of considering this parameter, which is variable, as the value of mass which doesn't change, instead.

5.2.2 *New unit of measure for the kg mass*

With an email sent the 14th September 2017, addressed to the President of the Republic, to the President of the Council of the Ministers, the Miur, the Scientific Community (in particular INFN, BIPM and INRIM), I announced my studies on "The shape of solid bodies".
The Scientific Community, on the occasion of the 26th Conference on Weights and Measures held in Versailles the 20th May 2019, declared that new definitions of the fundamental quantities would come into force. On the definition of the kg mass, neither the fabric nor the shape of the sample considered were fixed. This means that two samples of 1 kg mass, or of every other value, with different shapes and fabrics (in theory two infinite ranges of values are possible), when placed on a scale

(*measuring the weight and not the mass*) do not weight the same.

This is proof that, despite the technological and scientific advance, the Scientific Community still mistakes the body mass for the body weight, thus making mistake after mistake.

5.2.3 *Inertia*

The Inertia adopted over two years ago by the Scientific Community, perpetrating the unjustified refuse to a scientific and democratic confrontation, breaking the Constitutional and Institutional commitments, entailed the last mistake on the definition of the kg mass, demonstrated in this study.

Moreover, this inertia is leading to a delay on the assessment of the historical discovery on buoyant bodies, partially led by Archimedes, and partially by Stevin and Galileo, considered concordant, when, instead, they aren't.

Galileo himself erroneously made the conclusion that the results of his studies agreed upon the studies of Archimedes. In his work, about the study of buoyant bodies (DISCOURSE ON BODIES THAT STAY ATOP WATER, OR MOVE IN IT), his conclusion was:
"This is sufficient, for the present work, having with the previous examples discovered and demonstrated, without going any further, as it possible to discuss furthermore; instead, if it hadn't been necessary to solve the previously expressed doubt, I would have stopped to what Archimedes has demonstrated in his first book "On bodies staying atop water", where on the whole the same conclusions are made and established, to wit, that solid less heavy than water stay afloat, the heavier sink to the bottom, the equal in heaviness float indifferently

41

everywhere, on the one condition of remaining under water.".

As already seen Archimedes' third proposition states something different:
- Third proposition - "Solid bodies with the same weight of the liquid, if submerged in it, dive in without descending below the surface of the liquid nor plunging to the bottom";

a completely different conclusion compered to Galileo's.

As previously seen, what is stated in academic textbooks isn't Archimedes' original version but the different version of Stevin and Galileo.

This scientific controversy between Archimedes' version and Stevin's-Galileo's has an historical relevance.

Is all of this unimportant?

For the Scientific Community Inertia, it appears so.

The Scientific Community has to win over its inertia as soon as possible. Against the previous contrasting versions, based on theorical models, my historical and scientific discovery is shrieking, it is an *experimental* discovery, indeed, with which I established that there are infinite positions of balanced stability, no matter the depth, below the surface of water, "body hanging in water".

Why are the inertial mass media silent, too?

Altogether, filled with the spirits of Giorgio La Pira, Martin Luther King and Gandhi, we will contribute as soon as possible to the growth of the Scientific Community and of the society.

6.1 Didactics

Postulates on the study of nature, in particular of Physics, when used, may have the same significance that they have in geometry.

These postulates need to be provided with extreme simplicity, as if an approximation to describe certain aspects of nature, whose essence remains impenetrable. And so, it's uncoherent to derive principles and natural laws and least of all theories from these postulates.

6.1.1 *Archimedes*

The concepts of balanced or unbalanced stability, for resting and hanging bodies, as a theorical model, use the approximate postulate on the directions of the weight gravity (considered straight parallel lines).

From this approximation, nevertheless, it is impossible to derive any natural principle, such as the principle of the indifference in equilibrium, as a theorical model, used both for resting and hanging bodies: *nature isn't indifferent.*

The results of these theorical models are compared with the ones derived from experiments. For this reason, mental experiments, when used, need to be practicable and fit in with reality.

6.1.2 *Stevin and Galileo*

A patient and maieutic experiment, real or mental, on the free fall in liquids could have annihilated the postulate on the uniformity of the specific weight gravity (nowadays called density) of liquids, based on which there should be an indifferent equilibrium at any depth. If the body is suspended in water at a certain depth after a free fall, how was it possible to accept the fact that with any possible external cause, when the body is displaced atop or on the bottom, the body would be stable and accepted indifferently by water in any possible new position, *nature is not indifferent*: beware of prejudices.

Galileo, keeping intact Archimedes' postulate on the directions of the weight gravity, states his principle of inertia based on the indifferent equilibrium of resting bodies: this is not acceptable.

6.1.3 *Didactic textbooks*

It is correct and imperative that didactic textbooks would illustrate and explain the Laws of Physics with new and alternative methods.

It is highly unfair that textbooks illustrate and explain the Laws of Physics from which they derive unfinished results or contrasting results, as I have shown and demonstrated in my works.

6.2 Scientific Research

6.2.1 *Holistic vision*

Knowledge keeps up with Science and Philosophy.

It is high time not to transform Physics into Philosophy and vice versa.

Physics, Science and Philosophy need to remain close and rooted to the Society.

Everybody is able to understand everything.
Everybody is able to speak about everything.

There is the ether, no there isn't, there is the vacuum; a vacuum not so empty but a quantum full vacuum.

We don't know exactly what the ether is or what the vacuum is or what the quantum full vacuum is, either for the microscopic world and for the macroscopic world.

I know about this entity, whatever its name is, which is full of all the effects of all the celestial bodies (who and how many are they?). These effects are called light, rays, …

Human beings: body, mind, conscience?

Go on with the Didactics, with the Research and with the Speculation, but with humility and passion, always ready to a scientific and didactic confrontation, supported by a philosophical approach.

6.2.2 *Experiments*

Scholars, alone or in a group of people, study with a method that suits them best, using all the postulates that they hold to need and might also formulate their own theory and let the Scientific Community know about it.

But scholars need to fulfil the necessary experiments supporting their theory and with which the theory is built upon; until then the theory remains unknown to the Scientific Community.

For ages the principle of the philosopher Karl Popper has been in force, which states that every scientific theory needs to be falsifiable to be valid.

I consider this principle pointless and harmful for two reasons:
1) whoever is able to refuse a theory holding that theory to be non-falsifiable;
2) it asks someone else to demonstrate the inconsistency of the theory, and so swapping roles.

I, so, do not hold this principle of falsification by Karl Popper to be true.

A theory supported by experiments is law. Nonetheless, whoever proofs the groundlessness of the theory with experiments, may add, correct or refuse the theory.

I, so, wish for the Scientific Community to take into account this scientific and philosophical belief.

6.2.3 *The shape of solid bodies*

The experience derived from my scientific discovery "*The shape of solid bodies*", when perceived by the Scientific – Academic- Didactic Community, leads to the opportunity of the detailed study, *in an approximate point of view,* of many natural aspects that cannot be studied otherwise without the adequate tools.

One above all, from the study of "*The shape of solid bodies*", as already underlined, the "*shape of the containers of fluids*" is drawn with the possibility to study furthermore their behaviour, in particular of gasses. How Boltzmann's constant and Avogadro's number have affected the calculus?

This discovery will offer the opportunity of having more accurate measures in every field, in particular in quantum physics, to evaluate with more accuracy the values of body mass and electric charge of atomic particles.

6.2.4 *Historical Discovery*

The historical discovery related to the assessment of the presence of two versions on the study of buoyant bodies, illustrated above, is the perfect occasion to review under a differ light the past studies, in particular those without experimental proof, and so to inspire other scientists.

The studies on microcosm and macrocosm are welcome, and they are to be done simultaneously, or before, the consolidation of present studies.

For the love of the Research of the Truth and Knowledge, a *"Democratic and Scientific"* confrontation is needed: nobody can escape their obligations.

The experience of these past years is a reminder for the future history.

Not one.

Not a hundred.

But a thousand Odysseus.

No Giant.

With Socrates' art "no masters nor tutees"

For Peace and Quietness.
To Giorgio La Pira, Martin Luther King and Gandhi, who are the *mutual attraction transfigured.*

Acknowledgements

A special thanks to Salvatore Colombo together with whom I started, not by casualty, my evolution.
Thanks to my nephew Mauro Scala who has been throughout the reference point for my work.
Thanks to my son Pietro, who has drawn the pictures.